ELEMENTS OF GREAT MAINTENANCE MANAGEMENT

LUBRICATION: THE KEY TO MACHINE IMMORTALITY

JOEL LEVITT

Library of Congress Cataloging-in-Publication Data
Levitt, Joel 1952-
Lubrication- The key to Machine Immortality / by Joel Levitt Illustrated by Tatyana Nikolskaya
ISBN 9798387529566

1. Preventive maintenance
2. Tribology
3. Maintenance management
4. Lubrication

Lubrication: the key to Machine Immortality

2023

Springfield Resources, Havertown, PA 19083, USA

By Joel Levitt

Illustrated by Tatyana Nikolskaya

This book is dedicated to the millions of maintenance people who don't get any respect for lubricating their machines.

(c) 2023 Joel Levitt All Rights Reserved.

The book or parts thereof may not be reproduced, stored in a retrieval system or transmitted in any form without permission from the publisher.

What Other People Say About Elements of Great Maintenance Management

Sanya Mathura, Author & Managing Director at Strategic Reliability Solutions Ltd:
A captivating novel of the battle that reliability faces within most industrial plants! Joel expertly maneuvers the challenges that teams struggle with daily and provides quick and cost-effective methods of implementing changes that can positively affect your equipment's reliability. Overall, a great way to foster camaraderie amongst peers and get everyone on board with reliability.

Ramesh Gulati Author, Speaker, Reliability Sherpa, ReliabilityX:
Very interesting, very easy to read with alot of wisdom. Overall, alot of good information was presented in a very simplelanguage. So, the challenge is how can you make professionals read this?

Don Fitchett, President - Maintenance, Engineering Training Co., Nevada:
Although we had heard many of those perspectives in Joel's book from the stakeholders before, it is nice to have them in one presentation to get a better sense of the big picture. The 3D analysislab was super cool! The real-world examples andlearning process I watched (read?) play out was of great value to the story. Your book Joel is a real piece of art. You wrote it in such a way that it is relatable to everyone. It didn't hurt that I have a soft spot for superhero comics. :) Thanks for sharing.

Rolly Angeles, Consultant at RSA Reliability and Maintenance Consultancy Firm (Philippines):
Joel Levitt is one of the maintenance gurus I admire and have much respect for. I have purchased some of his books which I used for reference. There are many authors out there, which you will have a difficult time digesting the contents of the book. Joel has found a way of overcoming this by writing a book in an animated format where it is straightforward to understand.
If you arelooking for a smart way to improve your plant, I highly recommend this book by Joel Levitt, which you can finish reading in one sitting. I give this book two thumbs up.

Doc Palmer MBA, Author, trainer, Managing Partner Richard Palmer & Associates, Inc.:
Hey Joel, Thanks forletting me read this. You are one of my early career inspirations for helping people in maintenance. Along time ago, you sent me a cassette tape out of the blue that included "we remember Noah's flood because we just think about disasters, not the work of keeping them from happening." Many, many thanks

Bill Keeter CMRP, CRL, CRE , Reliability Engineering Trainer & Consultant:
Joel's books are a must-buy!

Alain Le Bon National Engineering and Maintenance Manager at Cheetham Salt:
I am really finding the books quite useful. I think they are good baseline educators and for shop floor. simple easy to understand messaging andlike thelight read, comic book approach which seems to work to audience. We definitely have a battle for reliability on our hands! I really want to engage the shop floor and these books are useful for that. one key initiative this year is the TLC aspect which i am combining with 6 senses (5 senses plus intuition) to utilize while engaging with TLC once small pilot area at a time across couple of our plants. Willlet you know how it goes!

Trent McJunkin, CMRM:
Everyone shouldlearn the basics this way! Be brilliant in the basics!

Michael Shorb Author Maintenance Manager, Lancaster, PA:
I've read a few of your books, and they've helped me over the years. I appreciate what you do. Over the years, I have recommended your books to all of my direct reports and seen them on the shelves of many maintenance professionals and a few plant managers. **Ed Gagnon:** This is another great publication, Joel. What a great novel. With the challenges we have with getting new people into the trades and current trades retiring we are pushing fewer people to do more. Maintenance education publications that you create, with the great graphics you use, engage a variety of teaching methods that work for a broad spectrum of readers.

Drago FrKovic, President Croatian Maintenance Society and Member of the European Asset Management Committee:
I really enjoyed the new book …once again you pointed how lubrication, tribology and maintenence work are important for "moviements of the world" Here in the countryside often peasants say the " one who lubricates is the one who drives".

Bill Closser, BS, MBA, President of NSG, Board of Directors Greater East Tennessee SMRP:
Joel has a great way of making what seems mundane into fun. His latest illustrated guide "graphic novella" Lubrication: The Key to Machine Immortality" is a quick fun read that teaches many of basics with easy-to-understand explanations of the principles of lubrication and its application practices. The "dad jokes" (as my kids would call them) made me chuckle but also made me remember some of the key points Joel was trying to make. This is a great tool to explain the importance of good lubrication practices to your maintenance staff (and some of your managers too). I highly recommend it.

Hasan Avdić, Ph.D., associate professor, Deputy Chairman Regional Sustainable Energy Transition Center:
There is alot of literature that deals with the study of lubrication and determining the condition of equipment based on testing lubricants in the phase of equipment exploitation. However, books such as this one, which deal with the issue of reliability in this way, are almost non-existent in practice. The book - "Lubrication - The key to Machine Immortality", authored by Joel Levitt, represents a huge contribution to the research of equipment lubrication in the exploitation phase. This graphic novel is intended for experts in practice, with the possibility of using it in classes, as supplementary literature, at faculties where the subject of Tribology and the basics of maintenance is studied at undergraduate and master studies. Only authors such as Joel Levitt, who possesses superior knowledge and vast experience in studying the subject matter, can write a work in this way.

Reynaldo Marquez CRL-Black Belt (CPMM):
Joel, I read your book twice; it is so to the point. I wish many other books were as engaging as this one. Congratulations!!!! I have no recommendations for improvement, and you really did an awesome job! May all your future books be as good or better . 👍 Cheers and success.

Agus Wahyudi from Balikpapan, East Kalimantan, Indonesia:
I Love this. The comic book titled Quality is not an Accident by Joel Levitt is genuinely exceptional. I have read it, and I found it to be an insightful and inspiring read."

BASIC LUBRICATION

EVERYTHING YOU ALWAYS WANTED TO KNOW ABOUT GREASE AND OIL, AND WERE AFRAID TO ASK

THREE LAWS OF MAINTENANCE

DON'T FORGET:

IF IT MOVES, AND IT IS SUPPOSED TO MOVE, **LUBRICATE IT.**

IF IT MOVES AND IT IS NOT SUPPOSED TO MOVE, **DUCT TAPE IT.**

IF IT DOES NOT MOVE, AND IT IS NOT SUPPOSED TO MOVE, **PAINT IT!**

TABLE OF CONTENTS

LUBRICATION, THE KEY TO MACHINE IMMORTALITY

THE LUBRICATOR ATTITUDE	8
WHAT MAKES OIL AND GREASE SLIPPERY?	14
FRICTION	15
LUBRICATION HISTORY LESSON	19
LUBRICATION MAKES MAGIC HAPPEN	21
VISCOSITY, POLARITY	23
TYPES OF LUBRICANTS	26
LUBRICANTS ARE BUSY WITH OTHER TASKS TOO	28
THE GREASE GUN, MAINTENANCE AND OPERATION	29
WHAT IS A GOOD LUBRICANT?	33
LUBRICANT PROPERTIES AND ADDITIVES	34
OIL AND GREASE ANALYSIS	36
WHAT CAN YOU DO TO IMPROVE LUBRICATION?	41
BE AN ADVOCATE: WHAT CAN YOU CONVINCE MANAGEMENT TO DO TO IMPROVE	44

THE LUBRICATOR ATTITUDE

OUT IN THE PRAIRIE OF A TYPICAL FACTORY FLOOR YOU ARE ON YOUR OWN.

HAVE GREASE GUN, WILL TRAVEL

THE TWINS ARE THE BEST LUBRICATORS THERE ARE. BEING THE BEST REQUIRES ATTENTION TO DETAILS AND THE RIGHT MENTAL ATTITUDE

SENSES ARE OPEN

THE TWINS HAVE A SPECIAL STATE OF MIND WHEN THEY APPROACH A MACHINE. THAT STATE IS CALLED OPEN. THEIR EYES AND EARS ARE OPEN, THEIR NOSE IS OPEN, THEIR HANDS ARE READY. THEY ARE TAKING EVERYTHING IN.

THEY LOOK FOR SUBTLE SIGNS. THEY ARE QUIET AND THOUGHTFUL. THEY SEE AND HEAR EVERYTHING. THEY MIGHT TOUCH THE MACHINE IN DIFFERENT PLACES TO SENSE HEAT OR FEEL VIBRATION.

THIS EXAMINATION DOESN'T TAKE LONG

THEY CHECK THEIR LUBRICATION OR PM SHEET AND VERIFY THE TYPE OF GREASE, THE NUMBER OF STROKES AND THE NUMBER OF AND LOCATION OF THE GREASE FITTINGS. THEY CLEAN THE FITTINGS AND GREASE AS WRITTEN. AFTER COMPLETION THEY CHECK OFF [DONE] AND MOVE TO THE NEXT MACHINE

YOU MIGHT NOT REALIZE THAT THESE FOLKS PLAY ANOTHER VITAL ROLE. THEY PROVIDE EARLY WARNING OF DETERIORATION, DAMAGE AND DEFECTS

THEIR THOUGHTFUL, TIMELY AND BRIEF INSPECTIONS SAVE MACHINES FROM FAILURE. WRITING UP DEFECTS IS ESSENTIAL. WE CALL THIS CORRECTIVE MAINTENANCE

LUBRICATION IS BASIC TO MAINTENANCE.
IF YOU WANT A CAREER IN MAINTENANCE, YOU WANT TO GAIN MASTERY IN LUBRICATION. THE EFFECTS OF IGNORANCE CAN BE EXPENSIVE AND DOWNRIGHT DANGEROUS. I'M REMINDED THAT A LONG TIME AGO I WORKED IN THE ENGINE ROOM OF A BIG FREIGHTER.
WE WERE SEEING SOME PROBLEMS WITH THE ENGINE.

WE LOOKED AT THE OIL AND FOUND LITTLE CHUNKS OF METAL. THE ENGINE WAS IN WORSE SHAPE THAN WE THOUGHT. WE COULD FIX THE ISSUE EXCEPT THAT REAL BAD WEATHER WAS COMING IN.

THE CAPTAIN TOLD US TO GET OUT OF THE HARBOR AND INTO DEEP WATER WHERE WE WOULD BE SAFER.

ENGINE TEMPERATURES WENT UP AND POWER WENT DOWN. EVERYONE IN THE ENGINE ROOM WAS SWEATING IT BECAUSE WE HAD TO HAVE POWER TO MANEUVER. WE BARELY MADE IT THROUGH.

THE SHIP HAD 3 WEEKS OF DOWNTIME TO FIX THE ENGINE AND ABOUT A MILLION DOLLARS WAS SPENT.

IT COULD HAVE BEEN OUR LIVES.

FRICTION IS THE FORCE THAT RESISTS THE RELATIVE MOTION WHEN RUBBING ONE BODY AGAINST ANOTHER

THE RESISTANCE TO MOTION IS FRICTION

YOU CAN FEEL FRICTION AS HEAT. FRICTION CONVERTS ENERGY (MOVEMENT) INTO HEAT

FRICTION INCREASES AS THE LOAD INCREASES.

- DIFFERENT MATERIALS CREATE DIFFERENT AMOUNTS OF FRICTION. THAT FRICTION IS MEASURED BY THE RELATIVE COEFFICIENT OF FRICTION.

- FRICTION DOESN'T CARE ABOUT AREA OF CONTACT. SO, A BRICK LAID FLAT AND ANOTHER STANDING ON ITS END WOULD CREATE THE SAME AMOUNT OF FRICTION

FLAVORS OF FRICTION:

STATIC - NOT MOVING
(HIGHER FORCE THAN DYNAMIC)
DYNAMIC - ALREADY MOVING
(LOWER FORCE THAN STATIC)

HOW DO YOU FIGURE OUT HOW MUCH FRICTION?

"LOOK OUT MATH COMING!"

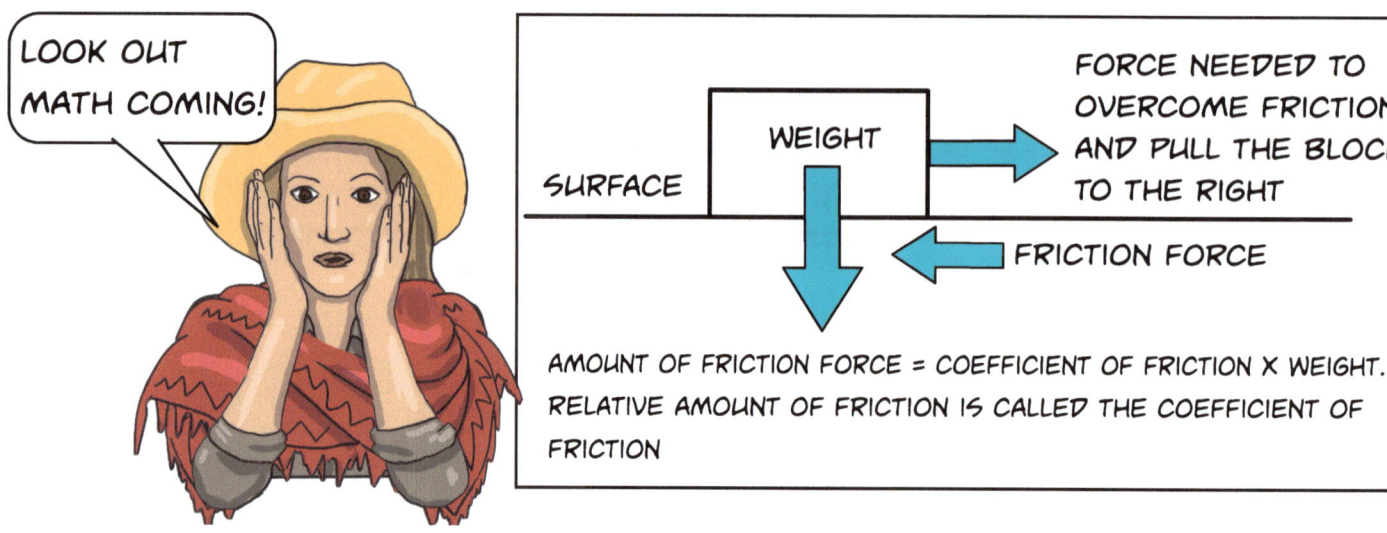

AMOUNT OF FRICTION FORCE = COEFFICIENT OF FRICTION × WEIGHT. RELATIVE AMOUNT OF FRICTION IS CALLED THE COEFFICIENT OF FRICTION

TYPICAL COEFFICIENTS OF FRICTION

SYSTEM	STATIC FRICTION μ
RUBBER ON DRY CONCRETE	1.0
RUBBER ON WET CONCRETE	0.7
WOOD ON WOOD	0.5
WAXED WOOD ON WET SNOW	0.14
METAL ON WOOD	0.5
STEEL ON STEEL (DRY)	0.6
STEEL ON STEEL (OILED)	0.05
TEFLON ON STEEL	0.04
SHOES ON WOOD	0.9
SHOES ON ICE	0.1
ICE ON ICE	0.1
STEEL ON ICE	0.4

WEIGHT	MATERIAL	COEFFICIENT	FRICTION FORCE
10#	SHOES ON WOOD	.9	9#
50#	SHOES ON WOOD	.9	45#
10#	SHOES ON ICE	.1	1#
20#	SHOES ON ICE	.1	2#
50#	SHOES ON ICE	.1	5#

THIS IS WHY FLOORS ARE MADE OF WOOD NOT ICE!

UNDERWRITERS LABORATORY (UL) SAFETY TEST FOR FLOOR WAX: THEY DRAG A BEAN BAG ACROSS THE NEWLY WAXED FLOOR AND IF PULLING FORCE IS TOO LOW THERE IS NOT ENOUGH FRICTION, AND THE WAX IS TOO SLIPPERY. NO UL APPROVAL!

WHERE DOES FRICTION COME FROM?

FRICTION COMES FROM THE INTERACTION OF IRREGULARITIES IN THE TWO SURFACES (AND SUCTION BETWEEN SURFACES).

ANY TWO BODIES, EVEN IF THEY LOOK SMOOTH, ARE MICROSCOPICALLY ROUGH. THEY ARE ROUGH, WITH SHARP, RUGGED PROJECTIONS, CALLED "ASPERITIES".

ALSO, MISALIGNMENTS IN THE TWO SURFACES CAUSE FRICTION.

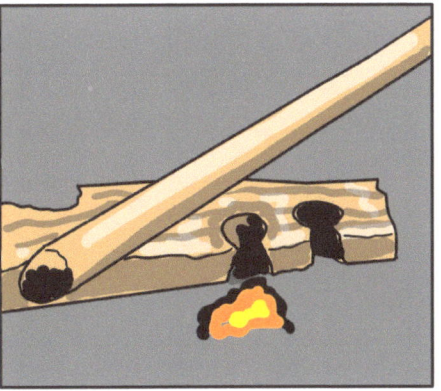

FRICTION IS A BEAUTIFUL THING AFTER A HARD COLD DAY IN THE FOREST WHEN YOU NEED A FIRE!

FRICTION CAN BE YOUR FRIEND (AND KEEP YOU OFF YOUR BOTTOM!

FRICTION IS GOOD OR BAD DEPENDING ON WHAT YOU WANT

BAD FRICTION: WHEN TWO BODIES RUB AGAINST EACH OTHER, THE ROUGHNESS SHEARS OFF PARTICLES AND CREATES HEAT

THE FRICTION WORE AWAY THE BEARING COMPLETELY

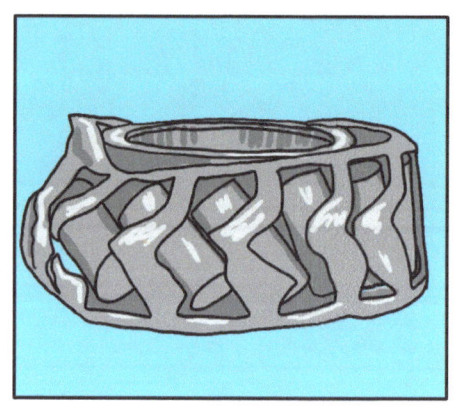

BEARING DESTRUCTION FROM HEAT CAUSED BY FRICTION

GOOD FRICTION: A BIT OF FRICTION MAKES THE WORLD GO AROUND. WITHOUT IT - TIRES, MATCHES, BOLTS, SHOES, KNOTS WON'T WORK.

FRICTION = HEAT IGNITES MATCH HEADS

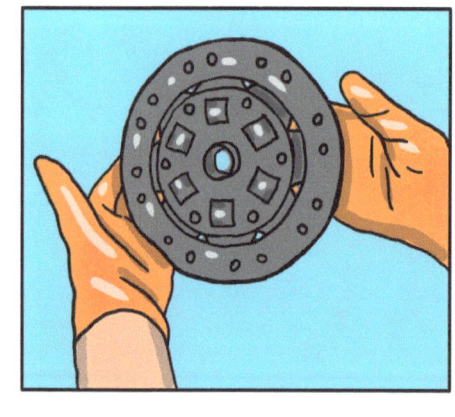

CLUTCH PLATES NEED FRICTION

YOU NEED NO FRICTION BETWEEN YOU AND THE SLIDE TO ENJOY THE RIDE

LUBRICATION HISTORY LESSON

1490 LEONARDO DA VINCI STUDIED THE PROBLEM OF FRICTION. ACROSS HORIZONTAL AND INCLINED PLANES, ALONG WITH WEAR ON SLIDE BEARINGS. THESE INVESTIGATIONS RESULTED IN HIS FIRST AND SECOND FRICTION LAWS.

FIRST EVIDENCE OF LUBRICANTS BACK TO 3500 B.C.E. AND THE ANCIENT EGYPTIANS AND SUMERIANS. THE WHEELED CARTS AXLES WOULD BECOME CHARRED BECAUSE OF HEAT. THEY USED A VARIETY OF LUBRICANTS, BITUMEN, ANIMAL AND VEGETABLE OILS, AND WATER, TO REDUCE FRICTION.

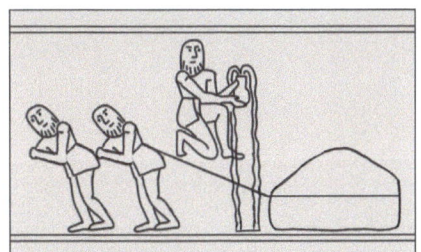

17TH CENTURY BC OLIVE OIL IS USED AS A LUBRICANT TO MOVE LARGE STONES AND OTHER HEAVY OBJECTS

17TH & 18TH CENTURY FRENCH PHILOSOPHER JOHN THEOPHILUS DESAGULIERS DESCRIBED THE EFFECT OF COHESION AND ADHESION ON FRICTION.

2600 B.C.E LUBRICATION WAS FOUND IN A SLED WHEEL THAT BELONGED TO AN EGYPTIAN PHARAOH

330 B.C.E., DIADES, A GREEK ENGINEER, DEVELOPED A ROLLER BEARING MECHANISM TO SUPPORT WARSHIP BATTERING RAMS. IT USED TRACES OF BITUMINOUS SUBSTANCES

8TH CENTURY WHALE OIL WAS USED IN LUBRICATING RUDDERS AND PULLEYS ON SHIPS

IN **1687**, SIR ISAAC NEWTON, DEFINED THE TERM "VISCOSITY," WHICH CENTERED ON THE BELIEF THAT FRICTION HAD BOTH MOLECULAR AND MECHANICAL CAUSES.

I SAW A 2000-YEAR-OLD OIL STAIN. IT MUST HAVE BEEN ANCIENT GREASE.

 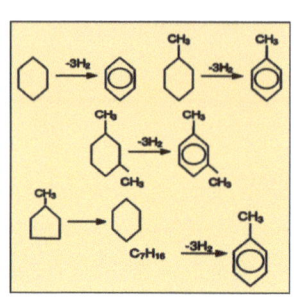 

1872. McCoy invented an automatic lubricator that applied oil through a drip cup to locomotive and ship steam engines. The term "the real McCoy" came about because engineers did not want low-quality copycat versions, so they would ask if it was "the real McCoy."

2020s February Orchid is a flower from China. It is a good bio-renewable component – could start a lubricant revolution

1845 Crude oil was used as a lubricant in a cotton spinning mill in Pittsburgh, Pennsylvania.

1966 The Jost Report included the first mention of the term "tribology." (the science of friction, wear and lubrication)

1850s The first oil well is drilled in Titusville, Pennsylvania, in 1859.

1930s-1940s, Additives prolong the performance and service life of automotive engine oils. The beginning of systematic oil analysis.

1950s Synthetic lubricants are developed, for use in the aviation and aerospace industries.

1990s Modern hydroisomerization technologies (transforming wax molecules into high-quality base oil) become widely used.

1970s Hydroprocessing technologies dramatically improve base oil purification and performance

BUT LUBRICATION MAKES MAGIC HAPPEN!

A BIT OF GREASE KEEPS THE BODIES FROM TOUCHING, SO THEY SLIDE BY EACH OTHER. THE VISCOSITY HOLDS THEM APART

THIS FLOATING ACTION LOWERS FRICTION DRAMATICALLY. HEAT AND WEAR DEBRIS PLUMMET. ENERGY (ELECTRICITY) NEEDED, PLUMMETS TOO

HOW DO LUBRICANTS LIKE OIL AND GREASE WORK?

FIRST: VISCOSITY HOLDS THE TWO MATERIALS APART

SO THAT ONE BODY "FLOATS" OVER THE OTHER

HIGH VISCOSITY = STRONG
LOW VISCOSITY = WEAK

GENERALLY, FOR THE SAME OIL:
HIGHER VISCOSITY = COLD = OIL IS THICKER = STRONGER FORCE
LOWER VISCOSITY = WARM = OIL IS THINNER = WEAKER FORCE

"SINCE MY GIRLFRIEND STARTED WORKING AT THE GREASE FACTORY.... IT'S BEEN REALLY HARD TRYING TO GET HOLD OF HER."

SECOND: OIL MOLECULES ARE NON-POLAR AND SLIDE BY EACH OTHER EASILY

OIL MOLECULES ARE SLIPPERY BUGGERS

POLAR LIQUIDS ARE LIKE MAGNETS. NORTH ATTRACTS SOUTH BUT REPEL ANOTHER NORTH. POLAR LIQUIDS DON'T MOVE AS FREELY. OIL IS NON-POLAR AND AVOIDS THIS PROBLEM.

OUR FRIEND VISCOSITY STILL HOLDS THE TWO MATERIALS APART

THESE POLAR MOLECULES BEAR SOME EXPLAINING

HONEY IS PRETTY VISCOUS. WHY DON'T YOU USE THAT FOR A LUBRICANT? IT TASTES GREAT TOO.

HONEY MOLECULES ARE POLAR. THEY GET CAUGHT ON EACH OTHER'S CHARGES AND DON'T SLIP BY EASILY

GUESS WHAT?

FACT OF THE DAY: OIL + SOAP = GREASE

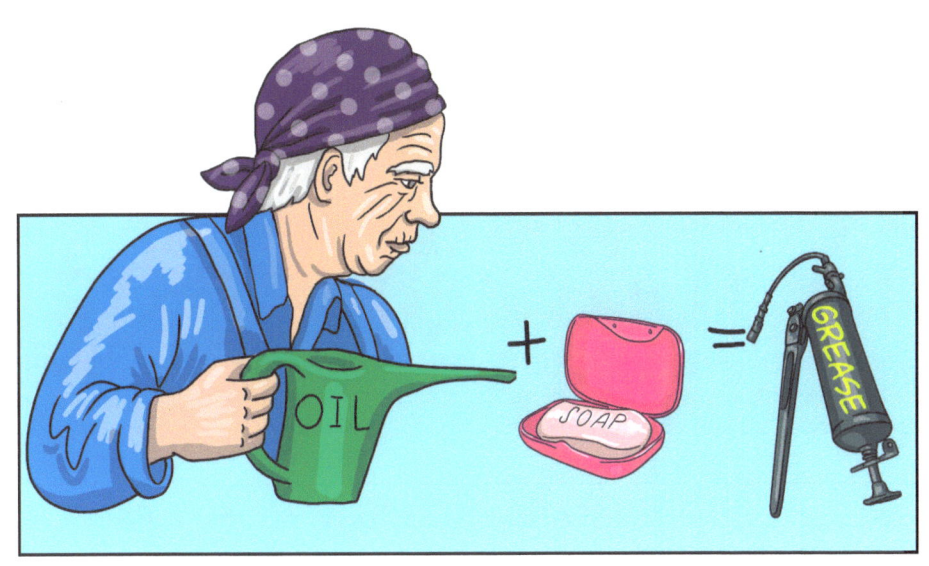

NO, NOT THIS GREASE!

IN CASE YOU WERE ITCHING TO KNOW: AN EXPERT IN LUBRICATION IS CALLED A TRIBOLOGIST

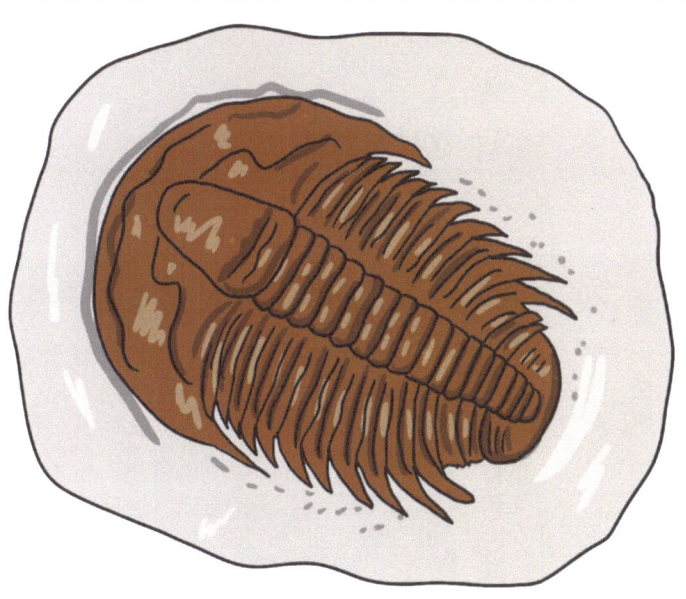

"I ABSOLUTELY ADORE THE TRILOBITE FOSSIL & WOULD LIKE TO HAVE ONE AS A PET!"

TRIBOLOGY IS THE STUDY OF SCIENCE AND ENGINEERING OF INTERACTING SURFACES IN RELATIVE MOTION. IT INCLUDES THE PRINCIPLES OF FRICTION, LUBRICATION AND WEAR. FIRST MENTIONED IN THE JOST REPORT OF 1966

THIS IS A FOSSIL OF A TRILOBITE FROM THE EARLY CAMBRIAN PERIOD (521 MILLION YEARS AGO) WHICH HAS NOTHING TO DO WITH TRIBOLOGISTS.

TRIBOLOGISTS MAKE THREE TYPES OF LUBRICANTS FOR THREE TYPES OF SITUATIONS

FLUID FILM – INSERTING A FLUID FILM THAT SEPARATES SLIDING SURFACES

VISCOSITY HOLDS THE SURFACES APART

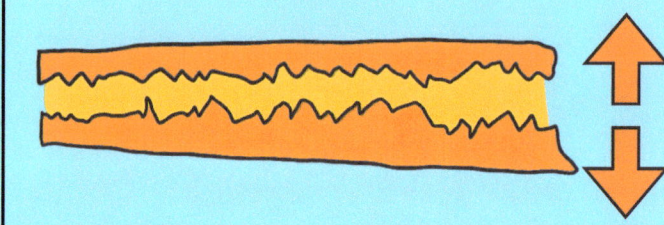

BOUNDARY – WHERE SURFACE FRICTION IS BASED ON PROPERTIES OTHER THAN VISCOSITY. COMMON DURING THE STARTING AND STOPPING OF MACHINES WHERE CONTINUOUS FILM OF LUBRICATION CANNOT PERSIST

SURFACES TOUCH, VISCOSITY FAILS TO HOLD THE SURFACES APART. LUBRICANT IS SUPER STRONG AND STICKS TO SURFACES

BOUNDARY LUBRICATION

SOLID – WHEN NORMAL LUBRICANTS DO NOT HAVE ENOUGH RESISTANCE TO HIGH LOADS OR TEMPERATURES, SOLID LUBRICANTS CAN BE USED. SOMETIMES METALS MAY EVEN BE USED AS LUBRICANTS.

EXAMPLES OF SOLID LUBES THAT ARE SUPER STRONG AND SLIPPERY

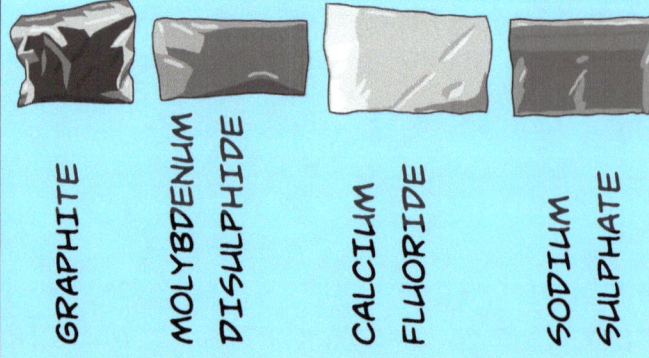

GRAPHITE | MOLYBDENUM DISULPHIDE | CALCIUM FLUORIDE | SODIUM SULPHATE

> ABOUT A MONTH BEFORE HE DIED, WE COVERED MY UNCLE'S BACK IN GREASE. HE WENT DOWNHILL FAST AFTER THAT.

HOLD YOUR HAT!
THIS IS WHAT LUBRICANTS LOOK LIKE HOLDING TWO SURFACES APART.

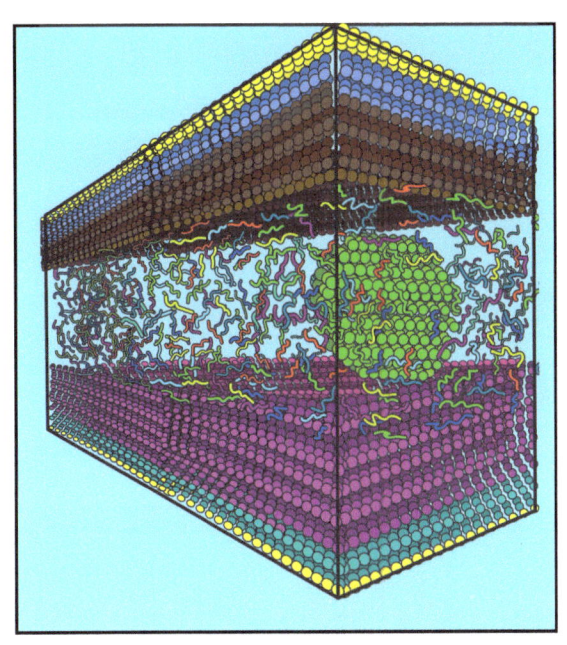

HIGH TECH! MOLECULAR MODEL AND SIMULATION OF LUBRICANT BETWEEN SLIDING SOLIDS. THIS IS DEEP SCIENCE. THIS IS THE WORK OF DR. SHAOPING XIAO, FROM THE UNIVERSITY OF IOWA

OIL HAS OTHER USES. IT IS USED TO REMOVE HEAT AND GET RID OF DIRT

OIL TRANSFERS HEAT AWAY FROM MOVING PARTS
OIL CATCHES AND TRANSPORTS DEBRIS, DIRT, AND CONTAMINATION AWAY FROM SURFACES THEY COULD HURT TO A FILTER THAT REMOVES THE BAD STUFF

ASK THIS QUESTION: "IS YOUR LUBRICANT REDUCING FRICTION, REMOVING HEAT AND GETTING CLEAN?"

MORE JOBS OF LUBRICANTS

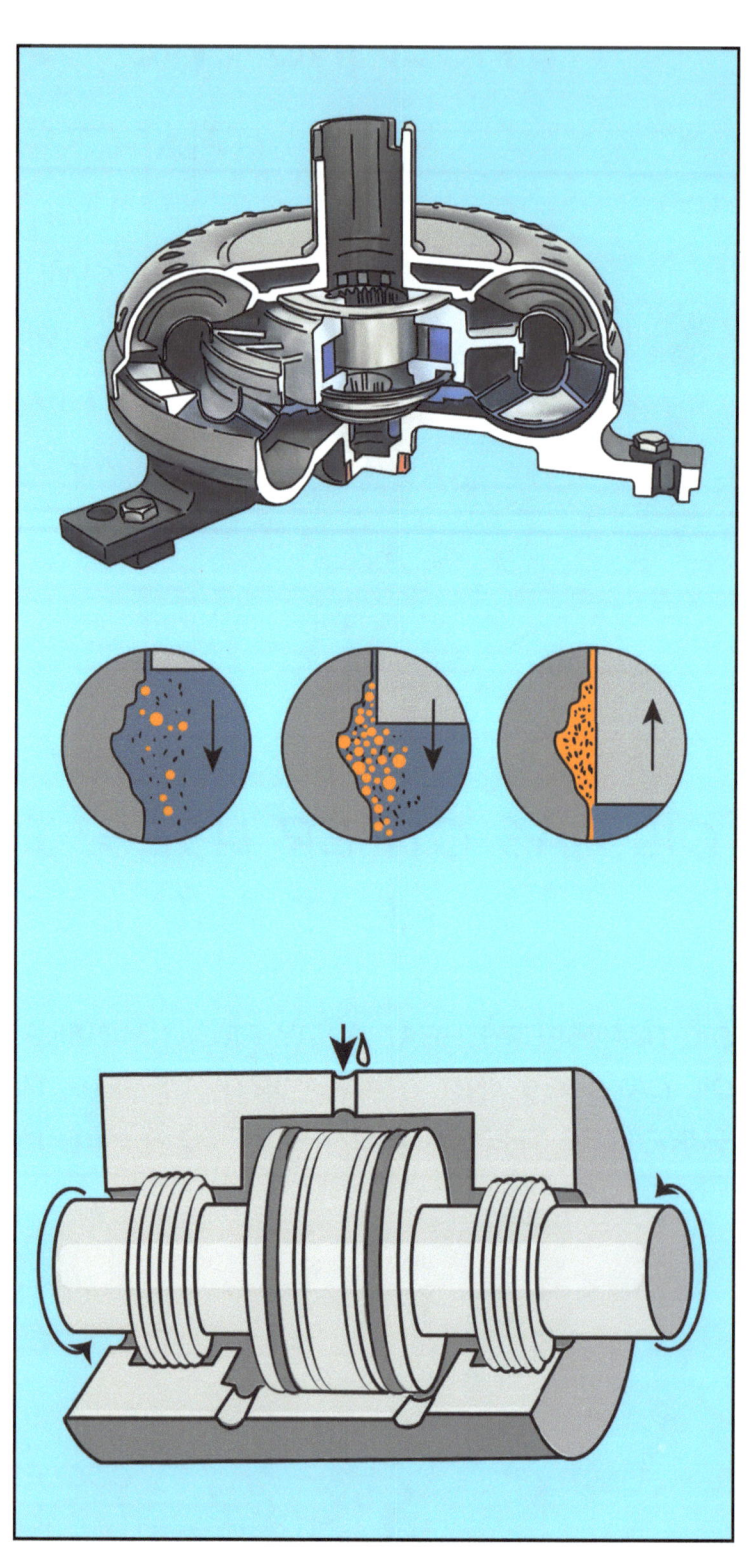

TRANSMIT POWER

LUBRICANTS KNOWN AS HYDRAULIC FLUID ARE USED AS THE WORKING FLUID IN POWER TRANSMISSION.

PREVENT CORROSION

MANY LUBRICANTS ARE FORMULATED WITH ADDITIVES THAT EXCLUDE MOISTURE TO PREVENT CORROSION AND RUST.

SEAL FOR GASES

LUBRICANTS WILL OCCUPY THE SPACE 2X BETWEEN MOVING PARTS THROUGH THE CAPILLARY FORCE, THUS SEALING THE CLEARANCE.

HOW MANY FARMERS DOES IT TAKE TO GREASE A COMBINE?
...ONLY TWO, IF YOU FEED THEM IN REAL SLOW.

THE GREASE GUN IS OUR PRIMARY WEAPON AGAINST FRICTION. LEARN IT, CLEAN IT, LIVE WITH IT UNTIL IT IS A NATURAL EXTENSION OF YOUR ARM.

LEVER ACTION HEAVY DUTY

- HEAVY DUTY ZINC PLATED LEVER
- RUBBER GRIP
- CAST ALUMINUM HEAD W/BULK LOADER & BLEEDER
- CLEAR BARREL
- BARREL END COLORS
- HEAVY DUTY ZINC PLATED LEVER
- RUBBER GRIP
- UNIQUE LOCK LEVER LOCKS PLUNGER ROD IN ANY POSITION
- NON SLIP FINISH

10,000 PSI | CARTRIDGE LOAD | SUCTION LOAD | BULK LOAD | VARIABLE STROKE

THANKS TO SAEPRODUCTS.COM

TYPES OF GREASE GUNS

1. MANUAL (LEVER)
THE MOST COMMON GREASE GUN SOLD IN THE MARKET.

2. MANUAL (PISTOL GRIP)
THIS TYPE OF LEVER GREASE GUN ENABLES THE OPERATOR TO ONLY USE ONE HAND IN PUMPING GREASE.

3. PNEUMATIC (PISTOL GRIP)
COMPRESSED AIR IS USED. TRIGGER ACTIVATES GREASE PUMP (UNTIL THE TRIGGER IS RELEASED)

4. BATTERY (PISTOL GRIP)
THIS GREASE GUN IS LIKE THE PNEUMATIC GREASE GUN BUT USES ELECTRIC BATTERIES

LET'S GET DOWN AND DIRTY!

YOUR GREASE GUN WILL HAVE THE FOLLOWING PARTS:

- A BARREL WHICH ACTS AS THE MAIN BODY OF THE GUN AND HOLDS THE CARTRIDGE OR TUBE.

- A GREASE CARTRIDGE WHICH CONTAINS THE GREASE

- A LEVER/PUMP/TRIGGER USED TO PUMP THE GREASE

- THE HEAD OF THE GREASE GUN WHICH HAS VALVES AND CONNECTS WITH A COUPLER

- A COUPLER WHICH CONNECTS THE HEAD AND NOZZLE OR HOSE OF THE GREASE GUN.

- A SQUEEZE WILL PUSH THE GREASE THROUGH NOZZLE ONCE PRESSURE IS APPLIED.

- FOLLOWER ROD HELPS THE PLUNGER FOLLOW A UNIFORM PATH AS IT KEEPS PRESSURE ON THE BOTTOM END OF THE GREASE TUBE.

ABOUT 1,375 OLIVES ARE PRESSED IN ORDER TO MAKE 1 LITER OF OLIVE OIL, 8,435 SUNFLOWER SEEDS TO MAKE A LITER OF SUNFLOWER OIL... DON'T EVEN GET ME STARTED ON BABY OIL

PRACTICE MAKES PERFECT

STEPS IN USING THE GREASE GUN:

- TO LOAD THE GREASE GUN, ITS BEST TO PULL THE FOLLOWER ROD BACK FIRST; BEFORE THE HEAD OF THE BARREL IS REMOVED

- MOST GREASE GUNS WILL HAVE A LOCK TO HOLD THE ROD IN POSITION. ONCE THE ROD IS PULLED BACK, USE THE LOCK AND SECURE THE ROD.

- OPEN THE NEW GREASE CARTRIDGE AND INSERT IT IN THE BARREL OF THE GUN.

- REPLACE THE HEAD OF THE GUN BUT DO NOT TIGHTEN IT.

- RELEASE THE FOLLOWER ROD FROM THE LOCK AND PUSH IT SO THAT IT WOULD FORCES THE GREASE FROM THE HEAD TO REMOVE THE AIR. PUSH THE ROD UNTIL SOME GREASE COMES OUT FROM THE HEAD.

- WIPE CLEAN (ALWAYS A GOOD HABIT) AND ATTACH THE NOZZLE TO THE HEAD.

- TIGHTEN THE HEAD AND RELEASE THE AIR KNOB TO RELEASE THE EXCESS AIR.

- ATTACH THE COUPLER TO THE NOZZLE, AND YOU ARE ALL SET TO USE THE GUN.

MY WIFE STARED AT ME IN DISBELIEF AND CRIED, YOU'RE SHIRTLESS AND COVERED IN... OIL?! I CHUCKLED PROUDLY, "WELL, YOU'RE ALWAYS SAYING.... I NEVER GLISTEN!"
SHE SCREAMED, **"LISTEN!!** YOU NEVER **LISTEN!!**

TAKE CARE OF YOUR GUN AND IT WILL TAKE CARE OF YOU!

MAINTENANCE OF GREASE GUN

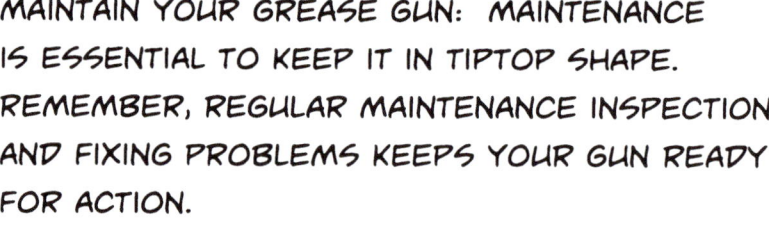

MAINTAIN YOUR GREASE GUN: MAINTENANCE IS ESSENTIAL TO KEEP IT IN TIPTOP SHAPE. REMEMBER, REGULAR MAINTENANCE INSPECTION AND FIXING PROBLEMS KEEPS YOUR GUN READY FOR ACTION.

CLEAN GUN IMMEDIATELY WITH A CLEAN CLOTH AFTER USE.

USE NOZZLE CAPS TO PREVENT DIRT OR IMPURITIES FROM GETTING INTO THE GREASE.

MAKE SURE THE COUPLERS, AIR NOZZLE, AND BARREL HEAD FIT AND DON'T LEAK

OVER-TIGHTENING WILL LEAD TO EXCESS AIR PRESSURE AT THE NOZZLE AND BARREL HEAD.

WHEN NOT IN USE, COVER YOUR GREASE GUN WITH A CLEAN CLOTH OR PUT IT INTO A BAG .

WHEN INSERTING A NEW CARTRIDGE, YOU CAN USE ISOPROPYL ALCOHOL FOR CLEANING THE BARREL. YOU WANT TO REMOVE ALL OIL OR DIRT BEFORE RELOADING.

AVOID MIXING OF OLD GREASE WITH THE NEW ONE. CLEAN OUT THE OLD GREASE INSIDE THE BARREL THOROUGHLY.

WEAR PPE SUCH AS GLOVES AND A FACE PROTECTIVE SHIELD OR SAFETY GLASSES.

MOST LUBRICANTS ARE INFLAMMABLE. DO NOT USE NEAR A FLAME. NO SMOKING NEARBY EITHER!

DOES ANYONE KNOW WHERE I CAN GET A LONGER DIPSTICK? MINE DOESN'T REACH THE OIL ANYMORE..

WHAT MAKES A GOOD LUBRICANT?

A GOOD LUBRICANT GENERALLY POSSESSES THE FOLLOWING CHARACTERISTICS:

- A HIGH BOILING POINT AND LOW POUR POINT (TO STAY LIQUID WITHIN A WIDE RANGE OF TEMPERATURES)
- APPROPRIATE VISCOSITY FOR THE JOB
- THERMAL STABILITY
- HYDROLYTIC STABILITY – A FANCY WAY OF SAYING DESTRUCTION OF OIL DUE TO PRESENCE OF WATER (WON'T SEPARATE)
- CORROSION PREVENTION
- A HIGH RESISTANCE TO OXIDATION
- LOW COST FOR THE APPLICATION
- LONG LIFE
- TRY TO HAVE AS FEW LUBRICANTS AS POSSIBLE

WHAT MAKES OIL BOIL? THE LETTER 'B'

LUBRICANTS ARE COMPLICATED!

Many additives are used to improve the performance of the lubricants. Modern lubricants might contain as many as ten additives, up to 30% of the lubricant.

Families of additives include:

- Pour-point depressants prevent clumping of waxes.
- Anti-foaming agents
- Additives allowing lubricants to remain viscous at higher temperatures.
- Antioxidants slow the rate of destruction of the oil
- Detergents ensure the cleanliness of contact surfaces.
- Rust inhibitors.
- Anti-wear additives
- Anti-scuffing additives that form protective films
- Friction reducers lower friction and wear

CHOOSING A LUBRICANT IS A DECISION USING TRIBOLOGY, MECHANICAL ENGINEERING AND MATERIALS SCIENCE. BETTER HAVE A BIG SLICE OF EXPERIENCE TOO!

ENGINEERING AND PURCHASING QUESTIONS

- ULTIMATELY, WHAT IS THE LONG TERM, MOST COST EFFECTIVE LUBRICANT FOR THE APPLICATION?
- WHERE DO YOU FIND THE BEST LUBRICANT?
- HOW DO YOU INCREASE MACHINE RELIABILITY AND REDUCE THE OVERALL NUMBER OF LUBRICANTS?
- HOW DO YOU FIND OUT THE TEMPERATURES AND PRESSURE PARAMETERS OF THE APPLICATION?
- WHAT PACKAGING SHOULD WE USE?
- IS THERE A SHELF-LIFE CONSIDERATION?

CHOOSING LUBRICANTS

SOURCES OF INFORMATION

- TRIBOLOGIST: HIRED BY YOUR COMPANY TO LOOK AT APPLICATIONS
- TRIBOLOGIST WHO WORKS FOR YOUR VENDOR TO LOOK AT YOUR APPLICATION.
- SKILLED AND EXPERIENCED SALES ASSOCIATE
- YOUR MECHANICAL ENGINEERING STAFF
- OEM-ENGINEERING STAFF
- PURCHASING DEPARTMENT

"YER LEAVING MONEY ON THE TABLE WITH SEAT OF THE PANTS LUBRICATION PROGRAMS!"

WITH A LITTLE PRODDING OIL AND GREASE WILL GIVE UP SECRETS

OIL ANALYSIS IS WHEN A LAB CHEMICALLY AND MECHANICALLY ANALYZES A LUBRICANT SAMPLE TO DETERMINE ITS CURRENT PROPERTIES, CONDITION, AND SPECIFIC CONTAMINATIONS

- TO BEGIN: TAKE A SAMPLE OF THE LUBRICANT AND SHIP IT TO A QUALIFIED LABORATORY.
- LABORATORY ANALYZES OIL FOR PROPERTIES (LIKE VISCOSITY) AND CONTAMINATION (LIKE WATER, METALS)
- READINGS ARE RECORDED AND TRENDED.
- REPORT TELLS MAINTENANCE FOLKS WHAT MAY BREAK DOWN, WHEN OIL SHOULD BE CHANGED AND IF THERE ARE LEAKS
- USUALLY THEY'LL EMAIL OR CALL IF DAMAGE IS IMMINENT
- BIG OIL USERS (LIKE BUS FLEETS OR CONSTRUCTION COMPANIES) SOMETIMES BUY THE EQUIPMENT AND HIRE A TRIBOLOGIST AND DO THEIR OWN OIL ANALYSIS.

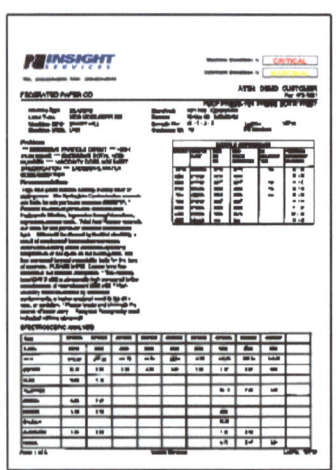
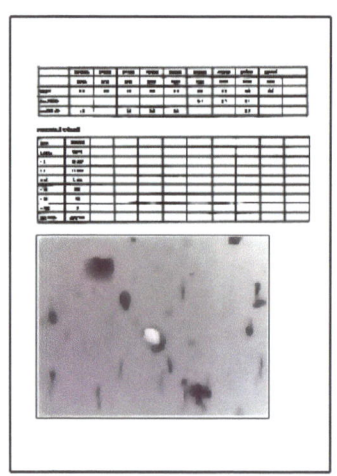
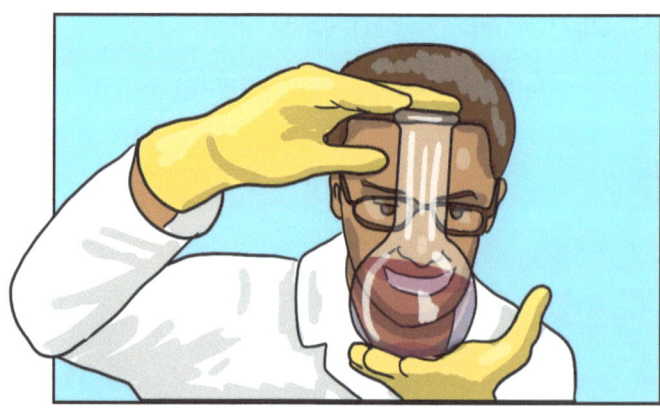

SCIENTISTS WILL ANALYZE YOUR OIL!

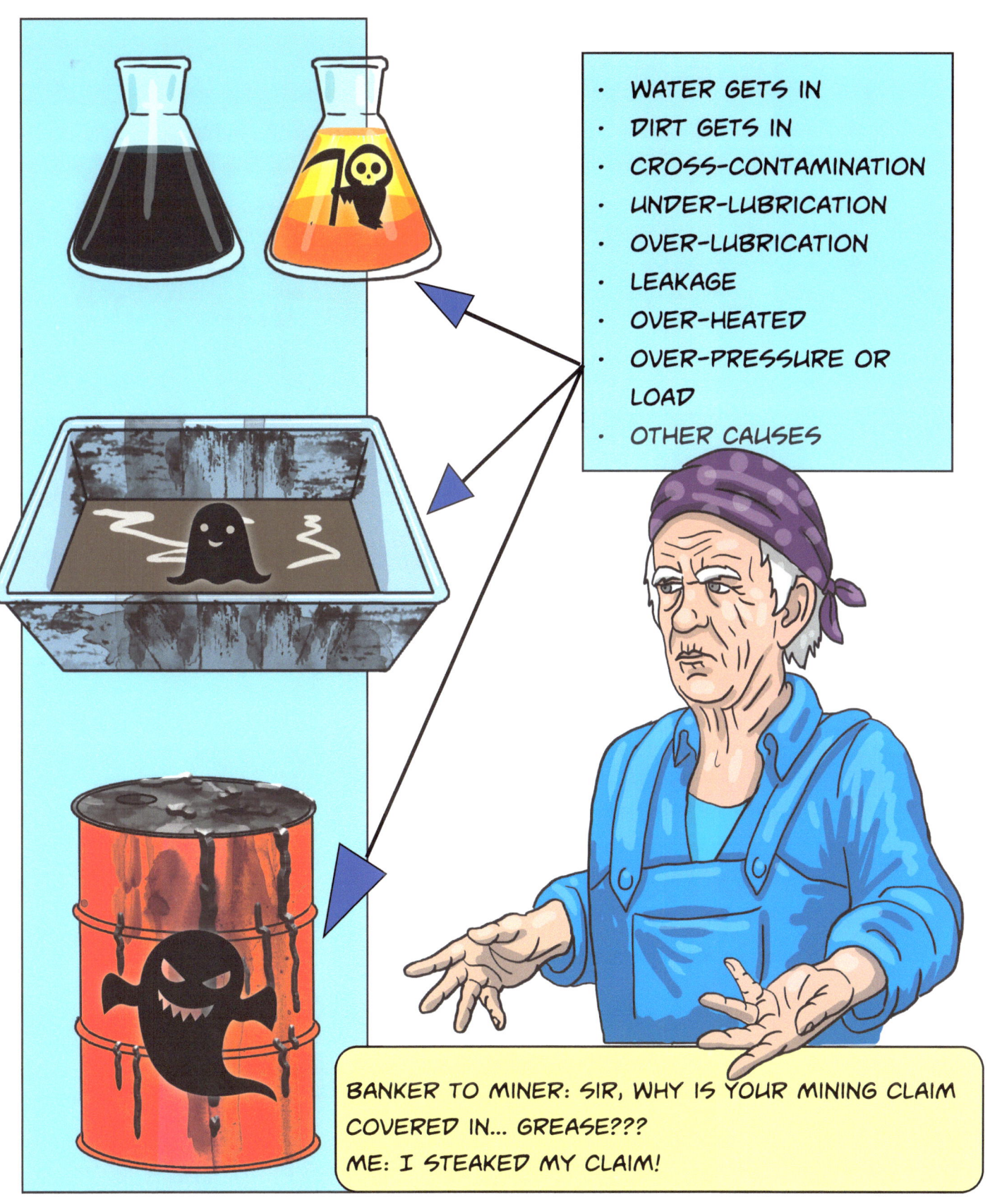

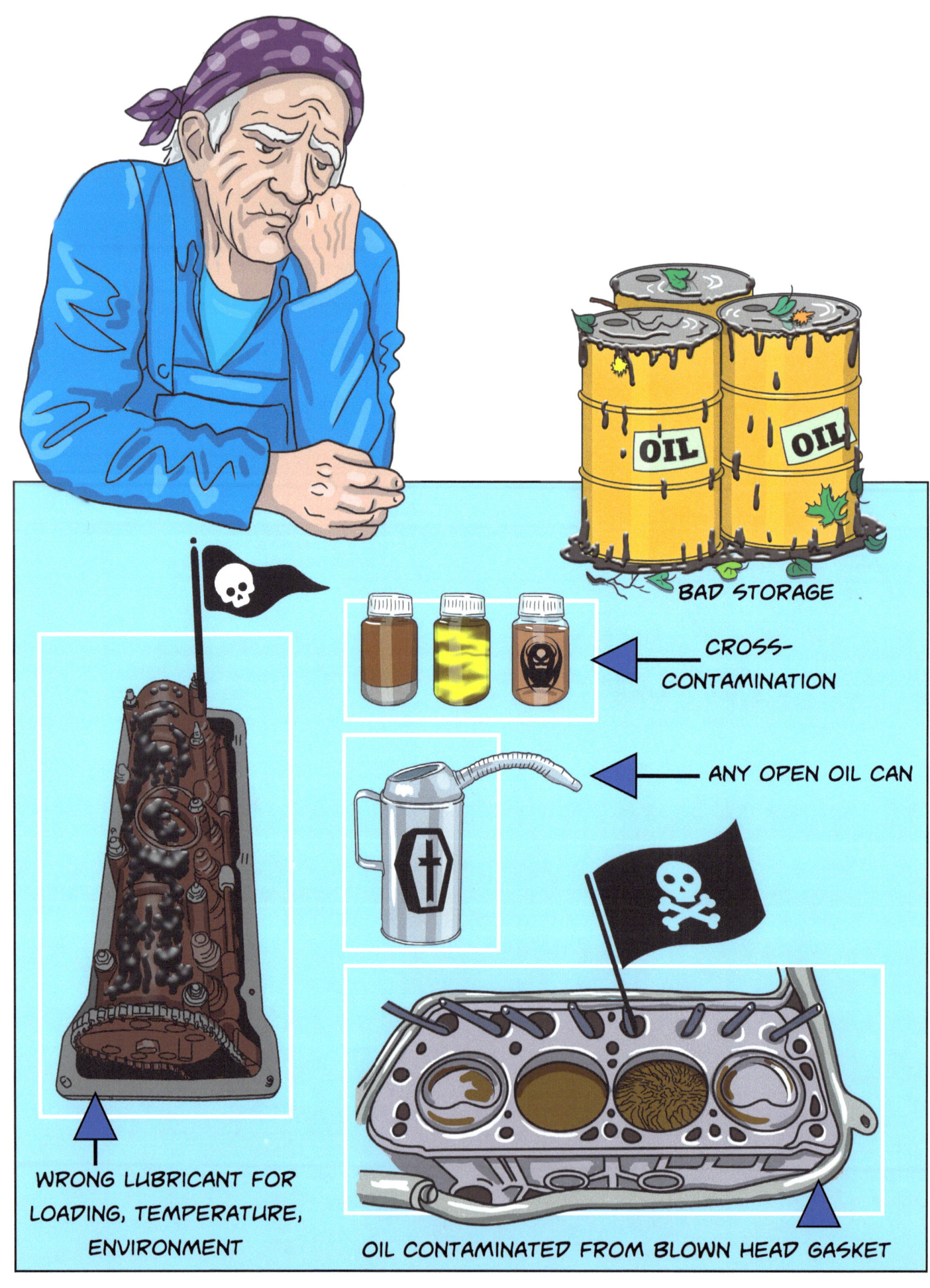

LEAKAGE CAN LEAD TO POLLUTION AND LOW OIL LEVELS AND CAN LET BAD CONTAMINATION IN

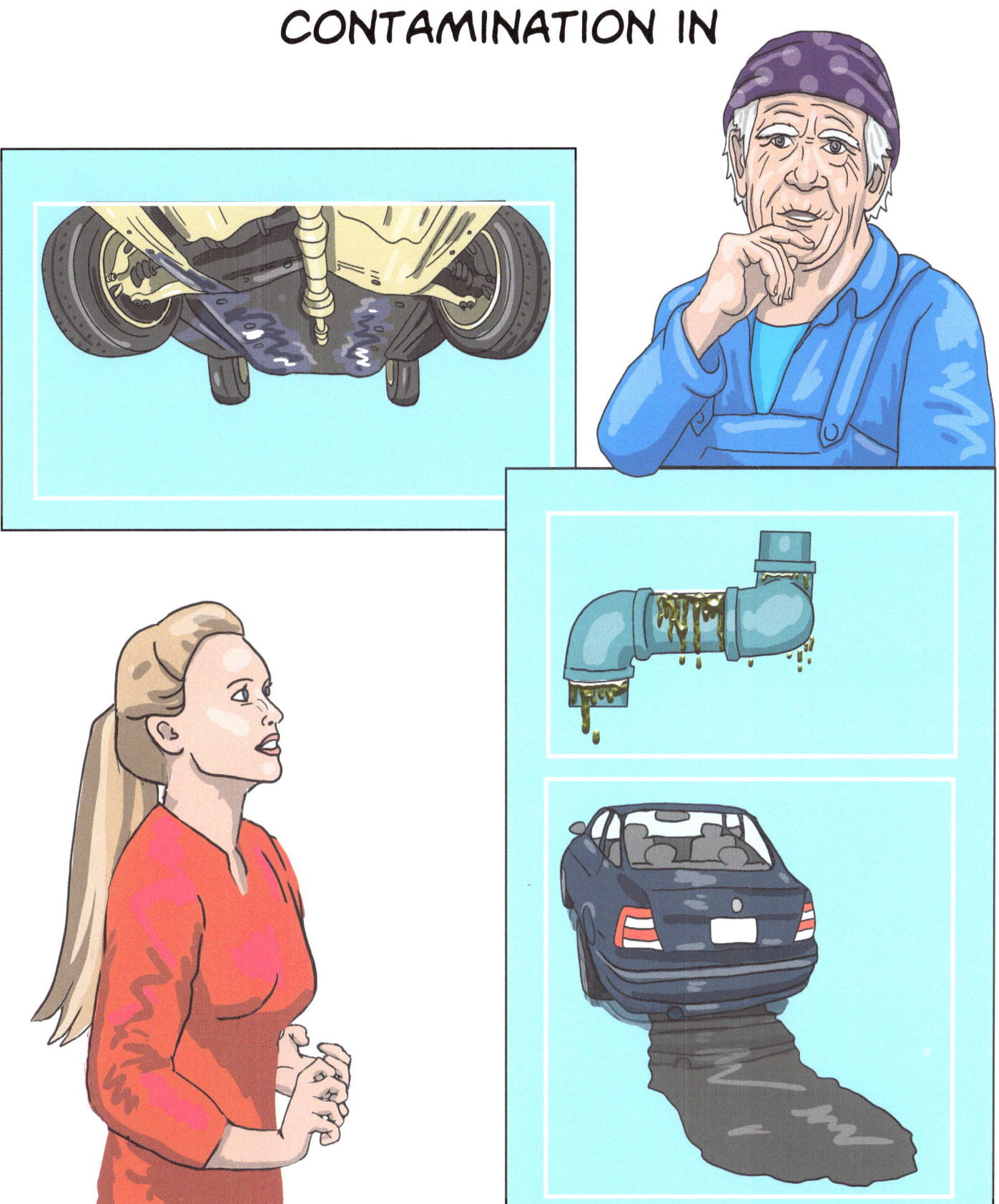

SO, WHAT CAN YOU DO?

YOU DO THIS:

- ALWAYS READ THE LUBRICATION TICKET OR PM FOR TYPE AND QUANTITY OF GREASE
- PAY ATTENTION TO THE MACHINE WHEN YOU LUBRICATE AND LOOK FOR ANY DEFECTS
- GREASE ALL THE FITTINGS (EVEN THE PAIN IN THE YOU KNOW WHAT ONES!)
- CLEAN ALL FITTINGS BEFORE LUBRICATION
- DON'T CLEAN THE FITTING AFTER GREASING UNLESS USING A CAP, A DOLLOP OF GREASE FORMS A SEAL
- BE SURE LUBRICANT CONTAINERS ARE ALWAYS CAPPED AND LABELED.
- ALWAYS USE THE SAME CONTAINERS FOR THE SAME LUBRICANTS, EVERY TIME

GREASE	APPLICATOR	RECEPTACLE
ABC GREASE #1		
RED STORAGE PAIL (BULK-GREASE PAILS OR GREASE-TUBE STORAGE AREA) →	RED GREASE GUN →	RED-PAINTED OR TAGGED GREASE POINT
XYZ OIL #3		
BLUE STORAGE RESERVOIR →	COVERED BLUE TRANSFER CONTAINER →	BLUE-PAINTED OR TAGGED RESERVOIR

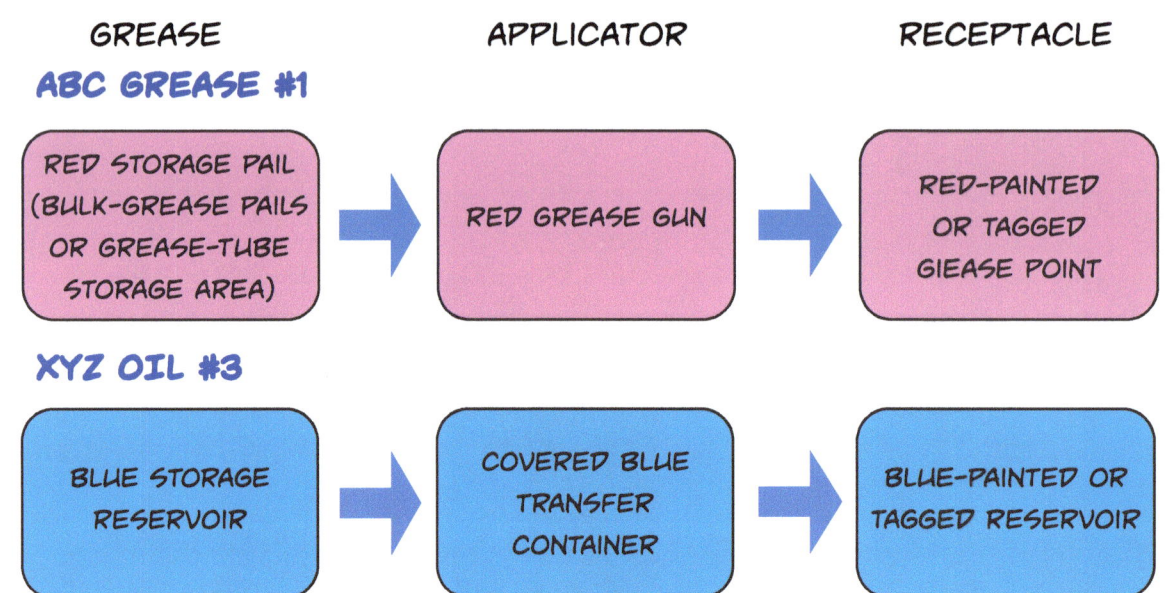

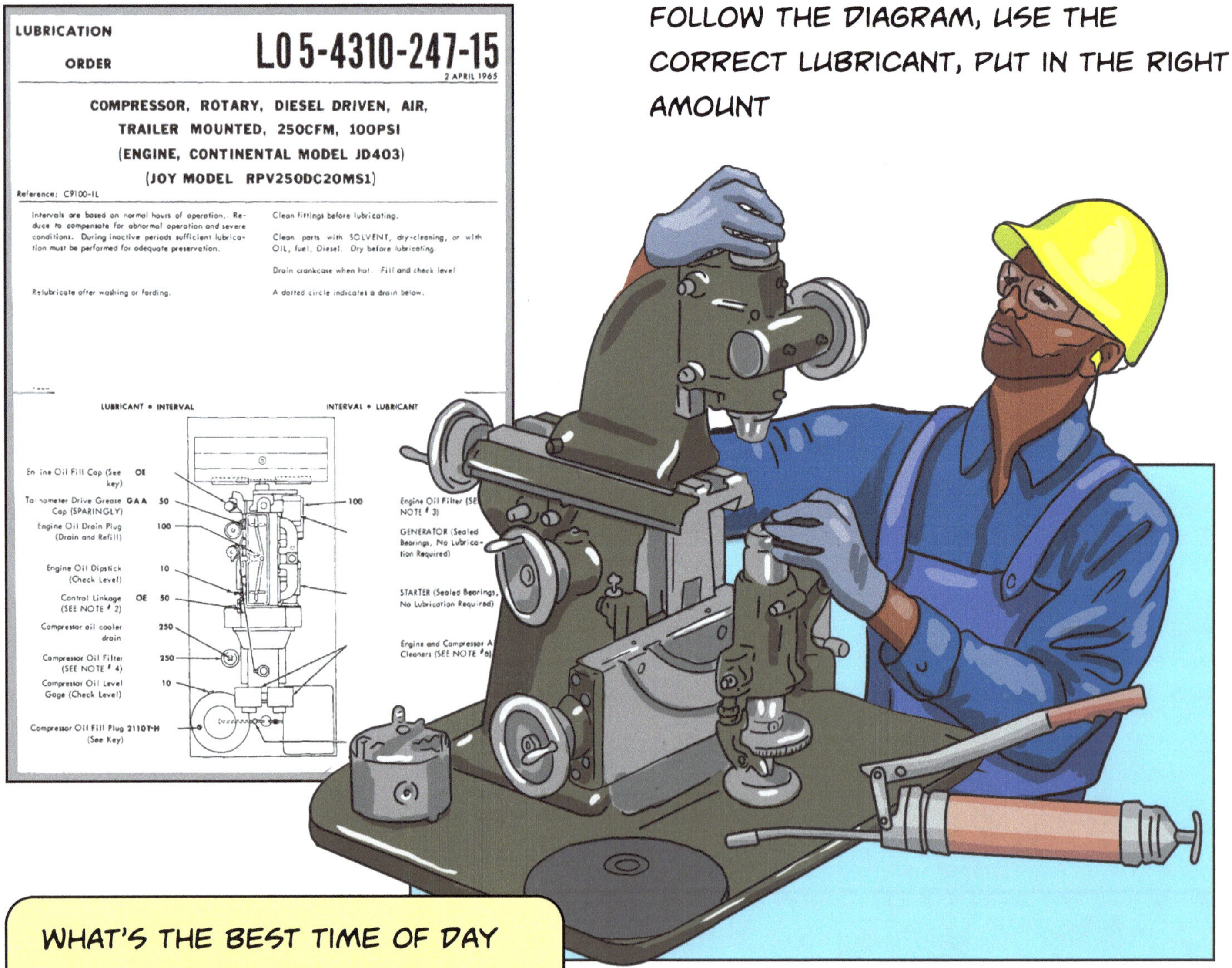

FOLLOW THE DIAGRAM, USE THE CORRECT LUBRICANT, PUT IN THE RIGHT AMOUNT

WHAT'S THE BEST TIME OF DAY TO WASH YOUR ENGINE? DAWN, IT'S TOUGH ON GREASE

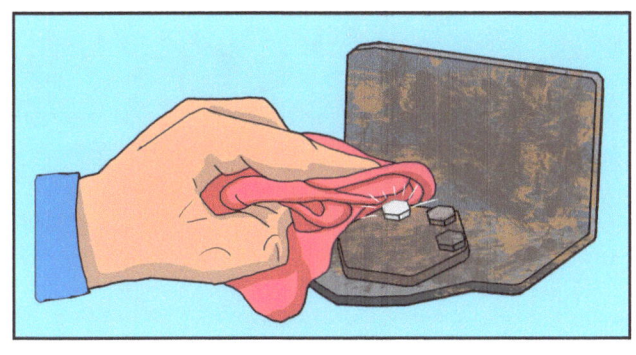

CLEAN BEFORE YOU LUBRICATE YOU DON'T WANT DIRT PUSHED THROUGH TO THE BEARING!

ENSURE THE SAME CONTAINERS ARE USED FOR THE SAME LUBRICANTS EVERY TIME AND ARE ADEQUATELY LABELED

COLORFUL, GEOMETRIC ID TAGS ALLOW EASY IDENTIFICATION EVEN IN LOW LIGHT.

COLOR CODED TO PREVENT CROSS-CONTAMINATION. THESE CAPS CAN BE SECURED TO MOST GREASE FITTINGS TO HELP PROTECT THEM FROM DAMAGE AND DIRT.

NO OPEN OIL CONTAINERS ENSURE LUBRICANT CONTAINERS ARE ALWAYS CAPPED

COOL FACTOID:
MOST COMMON RED--GREEN COLOR BLINDNESS AFFECTS UP TO 8% OF MALES AND 0.5% OF FEMALES. THE LABELING COMPANIES USE BOTH GEOMETRIC SHAPES AND COLORS TO DIFFERENTIATE LUBRICANTS

YOUR COMPANY SHOULD TAKE THIS SERIOUSLY!

TELL THE BOSSES WHY AND HOW TO IMPROVE LUBRICATION. YOU A HAVE A CRITICAL ROLE HERE. THE BOSSES DEPEND ON SUBJECT-AREA EXPERTS LIKE YOU TO MAKE GOOD DECISIONS. WITHOUT YOUR INPUT THEY WILL MAKE DECISIONS ON SHORT TERM COST ONLY!

SOME DISCUSSION ITEMS

- USE EXPLICIT VERSUS IMPLICIT TASKING IN PMS
- CONSIDER INVESTING IN AUTO-LUBRICATORS
- BE SURE THE QUANTITY AND FREQUENCY OF LUBE IS BASED ON **ENGINEERING** AND NOT JUST FOLLOWING THE LEADER OR THE PAST
- MAKE SURE THEY CHOOSE THE BEST LUBRICANT FOR THE JOB AND NOT JUST THE CHEAPEST
- CONSOLIDATE LUBRICATION TYPES TO A FEW AS PRACTICAL, CONSIDER MOVING TO HIGHER QUALITY LUBE THAT YOU CAN USE IN A LOT OF PLACES, THEN BUY IN HIGHER QUANTITIES
- RANDOMLY TEST INCOMING LUBRICANTS BY PERFORMING
- AN OIL ANALYSIS TO CHECK SPECS AND PRESENCE OF CONTAMINATION
- TRY USING ULTRASONIC GREASE GUNS
- STORE LUBRICANT INDOORS IN CLEAN, ADEQUATELY SIZED, GOOD LIGHTING, PROPER HANDLING EQUIPMENT LUBRICATION STORAGE AREA

EXPLICIT VERSES IMPLICIT TASKING ON PM SHEETS OR LUBE ROUTES

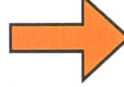 **IMPLICIT** TASKS ASSUME SIGNIFICANT PRIOR KNOWLEDGE

IMPLICIT
"HEY JACOB, GO PM THE COMPRESSOR." IMPLICIT TASKING IS ONLY SUITABLE FOR EXPERIENCED MAINTENANCE FOLKS.

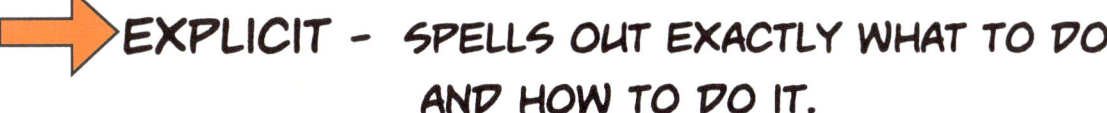 **EXPLICIT** - SPELLS OUT EXACTLY WHAT TO DO AND HOW TO DO IT.

GREAT IF YOU ARE NEW, DON'T DO IT OFTEN, WORK IN A COMPLEX ENVIRONMENT OR LUBRICATION IS INCIDENTAL TO YOUR DUTIES

- CONSULT PHOTO, DIAGRAM OR SKETCH (INCLUDED OR NEARBY)
- REMOVE CAPS
- WIPE GREASE FITTINGS WITH CLEAN RAG
- PUMP IN 1 SQUIRT OF RED LABELED (#3) GREASE
- RECAP. REPEAT 6X

AUTOMATED GREASERS:

PROS
GREAT FOR OUT OF THE WAY GREASE POINTS
GREAT IF YOU DON'T HAVE THE LABOR TO DO THE JOB

CONS:
LONGER INTERVAL BETWEEN VISITS FROM A HUMAN
COULD BE A PROBLEM IF IT RUNS OUT OF GREASE AND NO ONE NOTICES

AUTO LUBRICATION OILERS AND GREASERS
IT'S LIKE HAVING A 24/7 LUBRICATOR THAT LOVES WHAT IT DOES!

US $40

- CONTROL FREQUENCY
- CONTROL QUANTITY
- CONTROL RELIABILITY
- CONTROL CONTAMINATION
- CONTROL COST
- MULTIPLE POINTS
- ALERTS WHEN OUT OF LUBE, BROKEN OR CLOGGED

RANDOMLY SPOT CHECK INCOMING NEW OIL FOR CONTAMINATION AND SPECS

VENDOR DELIVERS OIL. OIL MEETS THE SPECIFICATIONS IN THE PURCHASE ORDER, OR DOES IT?

SEND SAMPLE TO INDEPENDENT OIL ANALYSIS LAB TO ANSWER QUESTIONS

DOES OIL MEET SPECIFICATIONS? IS IT CLEAN AND UNCONTAMINATED?

SOME COOL HI-TECH LUBRICATION GEAR: WE DESERVE THIS!

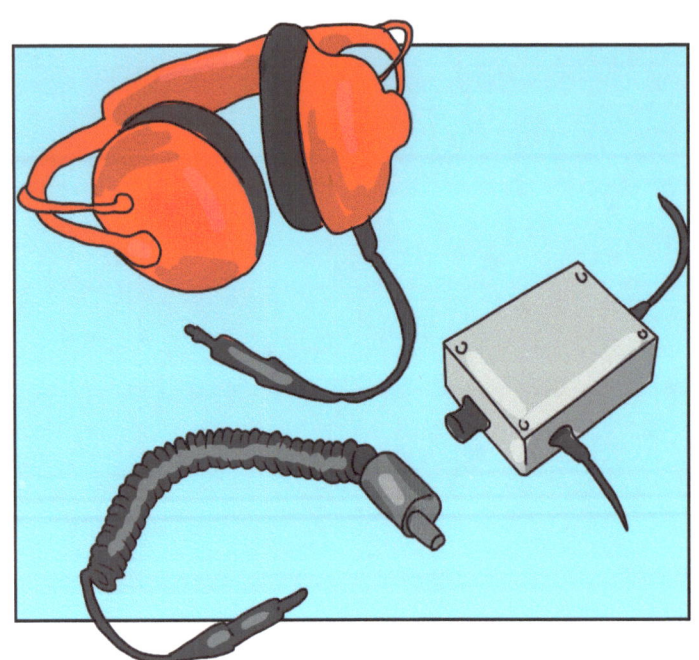

THE LUBRICATOR IS NOW ABLE TO LISTEN TO THE AMPLIFIED "ACOUSTIC" BEARING NOISE THROUGH THE HEADSET, HEARING PROBLEMS WITH LACK OF LUBRICANT OR TOO MUCH LUBRICANT.

MANY GREAT VENDORS. EXAMPLE: UE SYSTEMS ULTRAPROBE 201 ULTRASONIC GREASE CADDY

US $1900

WHY CHOOSE THIS?
- PROLONG THE LIFE OF YOUR EQUIPMENT
- SAVE ON LABOR
- SAVE ON OPERATING COSTS
- IMPROVE MAINTENANCE EFFICIENCIES

ALL THIS IS GOOD STUFF!
- CONFIRMS GREASE IS GETTING INTO A BEARING
- IDENTIFIES BEARINGS THAT ARE DAMAGED, UNDER LUBRICATED, OVERLOADED OR RUNNING AT EXCESSIVE SPEED
- ENSURING THAT GREASING QUANTITY AND FREQUENCY ARE ADEQUATE, TO ENSURE SERVICE LIFE IS ACHIEVED
- OPTIMIZES LUBRICATION INTERVALS
- DETERMINEbI QUANTITY OF GREASE REQUIRED
- GUARD AGAINST OVER-LUBRICATION (WHEN TO STOP PUMPING)

IS THE LUBRICATION STORAGE AREA CLEAN?
IS THE STOCK AREA ADEQUATE IN SIZE, LIGHTING, HANDLING EQUIPMENT FOR THE AMOUNT STORED?

GOOD SYSTEMS TO STORE AND DISPENSE WITHOUT CONTAMINATION AND TO AVOID MISTAKES. INADEQUATE HANDLING EQUIPMENT CAN CAUSE SMASHED TOES AND WRENCHED BACKS WHEN MOVING DRUMS.

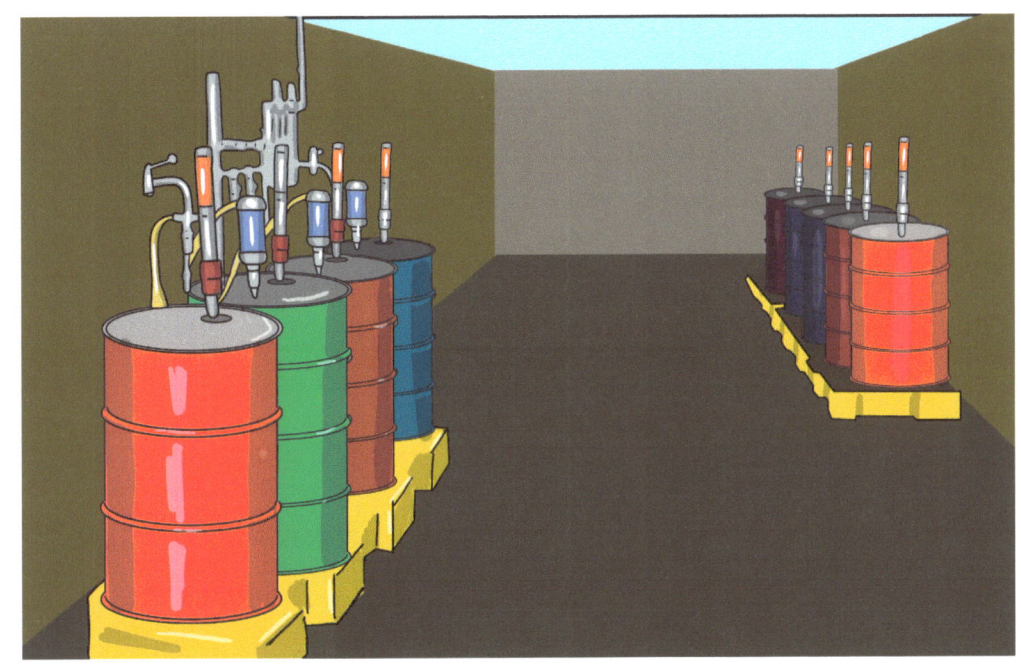

A RELATIVELY SMALL INVESTMENT HERE CAN IMPROVE RELIABILITY THROUGHOUT THE PLANT

IT MAY NOT SEEM TO BE TRUE, BUT LUBRICATION IS A 21ST CENTURY HIGH TECH ACTIVITY. IT IS THE CORE OF PREVENTATIVE MAINTENANCE AND THE LYNCHPIN OF EQUIPMENT RELIABILITY.

REMEMBER, IF IT MOVES, THEN LUBRICATE IT!

WE HAVE COME A LONG WAY FROM RUBBING TALLOW ON OUR AXLES

MY SPECIAL THANKS TO:
ED GAGNON
ELLIOTT GREENFIELD
HASAN AVDIC
JIM FITCH
PHILLIP SLATER
RAMESH GULATI
SANYA MATHURA MLE
WES CASH

THANKS FOR READING

THE END

About the Author

Joel Levitt CMRP, CRL, CPMM, Certified Change Management Professional is the President of Springfield Resources

He has 35 years in the field. He previously was Director of International Projects for Life Cycle Engineering and Director of Reliability Projects for Reliabilityweb.com.

Mr. Levitt is a leading trainer of maintenance professionals. He has trained more than 20,000 maintenance leaders from 3000 organizations in 42 countries in over 500 sessions.

He is a frequent speaker at maintenance and engineering conferences and has written 20 popular maintenance management texts and published over 200 articles.

In the past, Levitt served on the safety board of ANSI, Small Business United, National Family Business Council, and on the executive committee of the Miquon School. He is currently a member of AFE and Vice President of the Philadelphia chapter. He is a National Park Service volunteer Trail Maintenance Supervisor.

Books by Joel Levitt available from

http://www.maintenancetraining.com/bookshop

- Leadership Skills for Maintenance Supervisors and Managers
- Conversations in Maintenance
- Handbook of Maintenance Management
- Facilities Management - Maintenance of Buildings and Facilities
- Basics of Fleet Maintenance
- Managing Factory Maintenance
- Surviving the Spare Parts Crisis
- TPM Reloaded
- Maintenance Planning, Scheduling, and Coordination (with Don Nyman)
- Lean Maintenance
- The Complete Handbook of Preventive and Predictive Maintenance
- Managing Maintenance Shutdowns and Outages
- The Internet for Maintenance Professionals

Books of General Interest

- 10 Minutes a Week to Great Meetings
- 10 Minutes a Week for Great Time Management
- Sibling Revelry (with JoAnn Levitt and Marjory Levitt)

GRAPHIC NOVELS: ELEMENTS OF GREAT MAINTENANCE MANAGEMENT

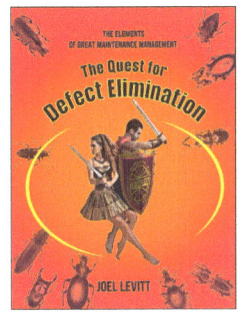

Quest for Defect Elimination

ISBN 9798627700458

Want more reliability? Then eliminate defects - NOW!

Quest for Defect Elimination is a full-color graphic novel that provides the basics of defect elimination and shows the quest to achieve fewer defects.

The Battle for Reliability, Fundamentals Win the Day

ISBN 9798598795286

The Battle for Reliability is a graphic novel teaching the critical need for all industries to focus on maintenance fundamentals to avoid unscheduled shutdowns from breakdowns. The battle for reliability shows that everyday failures are still killing productivity even with the current focus on hi-tech sensors and software.

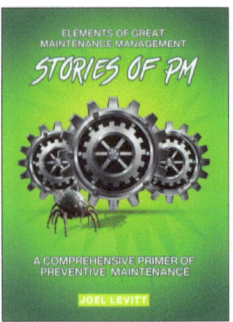

Stories of PM, A Comprehensive Primer of Preventive Maintenance
ISBN 9798496071154

Stories of PM tells the PM person the why, what, when, where, and even sometimes how to PM effectively. The stories are fun to read. Stories of PM tells the whole story of PM including the history. Included are about a dozen best practices disguised as the rantings of a bug.

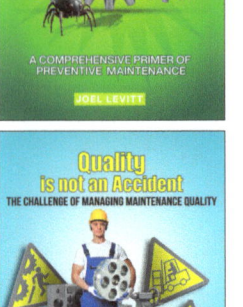

Quality is No Accident - The Challenge of Managing Maintenance Quality
ISBN 9798842723751

Quality issues are directly related to safety issues; at the core are mistakes. Quality is Not an Accident will directly help you reduce all types of mistakes. This full-color graphic novel is fun to read, and covers the basics of managing quality.

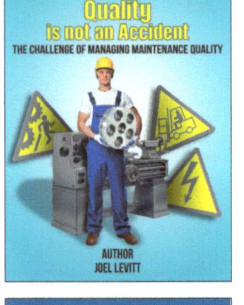

Lubrication - The key to Machine Immortality
ISBN 9798387529566

Lubrication is the simplest and most effective way to ensure the smooth running of your equipment. Yet, many people don't understand what they are doing, how to lubricate correctly, and why they are doing it. In the past, a senior person would mentor new employees and teach them what, why, and how.

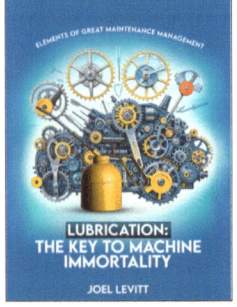

Maintenance Planning Cuts Waste and Frustration
ISBN 9798333645401

Maintenance Planning Cuts Waste, and Frustration provides your organization with a lighthearted, fun, reality-based graphic novel format training tool. Maintenance Planning starts with examples of a lack of planning, shows why there is so much resistance to planning, and then dissects and demonstrates techniques for effectively planning maintenance work.

CMMS - Friend or Foe
ISBN 9798311159500

Are you frustrated by your Computerized Maintenance Management System (CMMS)? This graphic novel breaks down the complexities of maintenance management, turning dry technical details into eye-opening, practical wisdom, the real purpose of CMMS—and how to make it work for you.

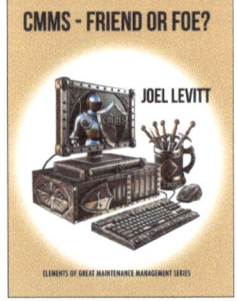

SIP – Small Improvement Projects
ISBN

Small Improvement Projects (SIPs) are short-term, action-driven initiatives aimed at eliminating defects, reducing maintenance, and improving reliability. They focus on quick, high-impact changes requiring minimal investment

http://www.maintenancetraining.com /bookshop

www.ingramcontent.com/pod-product-compliance
Lightning Source LLC
Chambersburg PA
CBHW051218220526
45473CB00003B/1079